# The Painter's Palette

## *A Theory of Tone Relations, an Instrument of Expression*

By Denman Waldo Ross

*Lecturer on Design in Harvard University*

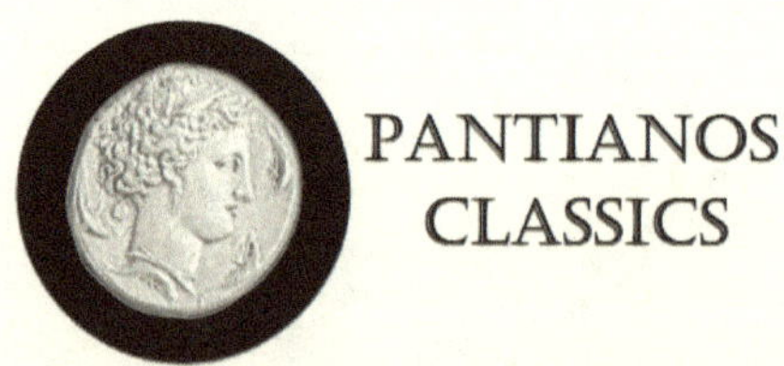


Published by Pantianos Classics

ISBN-13: 978-1-78987-515-7

First published in 1919

# Contents

## *Preface*

In taking up the practice of painting, it is a question what pigments to use, how to arrange them on the palette, and then how to use the palette. Following the example of most painters, we take certain pigments; blacks, browns, reds, yellows, greens, blues, and whites; and, putting them in a row, we proceed to mix them, two or more together, quite freely, following our visual impressions or the suggestions of the imagination. In so doing we hope to produce results and effects which will be appropriate to what we have seen or to the ideas we have to express. There is no principle or law in this procedure and the effects produced depend, in every case, upon the taste and judgment of the performer; who is, necessarily, as much concerned with the problems of the palette as he is with the problem of his Motif or Subject, whether it is well chosen; of his Design, whether it is consistent; or of his Drawing, whether it is true or expressive. It is so difficult to decide what to do, in view of the great number and variety of pigments, and the infinite possibilities of mixing them together. Most painters get over the difficulty by acquiring a habit and doing the same thing repeatedly, following the precedents of their own particular practice, whatever it happens to be.

Considering the Art of Music and the use of musical instruments, it seems that the musician has a great advantage over the painter in having a fixed scale of tones and definite rules for using it, - rules based on good precedents and representing the practice of recognized masters. Thinking of musical instruments and the laws of Counterpoint and of Harmony, the question comes up whether it may not be possible for the painter to convert his palette into an instrument of precision and to make the production of effects of light and color a well ordered procedure, - a procedure which everyone can understand and follow. If this is possible, the mind of the painter may be released from the problems of the palette and devoted, almost wholly, to the still more important problems of Subject, Design, and Representation. After more than twenty years given to the consideration of this question and to experiments in the use of set-palettes, I am fully persuaded that it is perfectly possible to make of the painter's palette an instrument of precision, - an instrument which will serve him both as a mode of thought and means of expression. He will then use his palette very much as the musician uses his voice or the violin or the piano.

Among the many palette-systems which I have devised and considered there are two which seem to me particularly interesting and promising. One of these systems may be described as the System of a Suitable Triad Repeated. The other may be described as the System of the Spectrum Band with Complementaries in Corresponding Values.

In the first system; a certain triad of colors, a triad in which there is a complementary balance, is repeated at equal intervals of the Scale of Values between the extremes of Black and White. There are four triads which I have used in this way: Red, Yellow, Blue; Orange, Green, Violet; Violet-Red, Orange-Yellow, Green-Blue; Red-Orange, Yellow-Green, Blue-Violet. There is another series of triads in which the colors are taken, not at equal intervals of the Scale of Values, but at unequal intervals. In the four regular triads above given, the colors are taken at the interval of the fifth. Taking the colors at intervals of the fourth, fifth, and sixth, we get another series of triads, with complementary balances, - a series which I have used and found extremely interesting. It was Mr. H. G. Maratta who first suggested them. The regular triads will be generally used; the others as they may be required. Taking the regular triad Red, Yellow, Blue, and repeating it five times between the extremes of Black and White, we get the palette indicated in the following diagram:

THE RED-YELLOW-BLUE PALETTE

| | | |
|---|---|---|
| White | White | White |
| Red | Yellow | Blue |
| Red | Yellow | Blue |
| Red | Yellow | Blue |
| Red | Yellow | Blue |
| Red | Yellow | Blue |
| Black | Black | Black |

In using this palette I have generally followed the rule of taking the colors on up-to-the-right or up-to-the-left diagonals, - mixing Red with a higher Yellow and Yellow with a higher Blue; or Blue with a higher Yellow and Yellow with a higher Red. By mixing the three colors, as they follow one another in one or the other of the two sequences, I get neutralizations and a perfect neutrality. Occasionally I have followed both sequences in the same design or picture. This System of a Suitable Triad Repeated is very fully described in my book, - On Drawing and Painting, published by Houghton Mifflin Company in 1912.

In this book, which is published as a supplement, I propose to describe and explain the System of the Spectrum Band with Complementaries in Corresponding Values. I have had this system in mind, in one form or another, for more than twenty years, and I have followed it, off and on, during all that time; following it for a while and then, preferring the System of a Suitable Triad Repeated, giving it up. During the past five years, however, I have used it almost exclusively; and, in its present form, it seems to me the better system of the two. It is much more logical in theory and much easier to follow in practice. The most serious difficulty will be found in the setting of the palette; in deciding what pigments to use, in preparing the tones and in getting the complementaries properly adjusted, to produce the required neutralizations and a perfect neutrality. It is only a master who can produce a satisfactory palette which will be the instrument of precision that we want. In the effort to produce such an instrument, however, the student will get a very valuable training which will give him, in some degree, the power of discrimination in tone-relations which, if he hopes to become a master, he must have. The student who follows carefully the indications and directions given in this book ought to be able, in due time, to set his palette correctly and to use it properly. After that, it will be a question whether he has any good reason for using it; whether he has anything to express that will be worth while.

I am not asking the painter to give up the System of a Suitable Triad Repeated if he has been using it with success. The system has its value and I can very well understand how it may be preferred; I have so often preferred it myself. I am simply asking him to consider the system which is described and explained in this book, and to give it a fair trial. He will then be in a position to decide which of the two systems he prefers. In my own experience and judgement, the System of the Spectrum Band with Complementaries in Corresponding Values is the better of the two.

# The Painters Palette

## Introduction - Terminology, Definitions

### TONES

With pigments and pigment-mixtures we are able to produce a great number and variety of effects, - effects of light and of color, which we call the tones of the palette.

In every tone thus produced there are two elements: there is the quantity of light in the tone that we call its value, and the quality of light in the tone that we call its color. The quantity of light in any tone and its color depend upon the pigments which we use in producing it. When two or more pigments are used, the tone depends, not only on the pigments but on the quantities and proportions of them that are used.

### VALUES

We have in black pigments (Blk); in the pigment Blue Black, for example; the least quantity and the lowest value of light. In white pigments (Wt); in Zinc White, for example; we have the greatest quantity and the highest value of light. All other quantities or values available for the Art of Painting lie between these extremes. At the half-point between the extremes we have a half-tone or Middle value (M); half way between Black and Middle we have a value which we call Dark (D); half way between Middle and White we have a value which we call Light (Lt). Between the five values thus defined we have four more which may be described, beginning with the lowest, as Low Dark (LD), High Dark

(HD), Low Light (LLt), and High Light (HLt). There are intermediates between these nine values which we are constantly using but they need no names, as we rarely, if ever, have occasion to speak of them or to write about them.

## THE SCALE OF VALUES

| | |
|---|---|
| White | (Wt) |
| High Light | (HLt) |
| Light | (Lt) |
| Low Light | (LLt) |
| Middle | (M) |
| High Dark | (HD) |
| Dark | (D) |
| Low Dark | (LD) |
| Black | (Blk) |

There are thirty-six contrasts which may be produced between one value and another of this scale; as follows:

## CONTRASTS OF THE SCALE OF VALUES

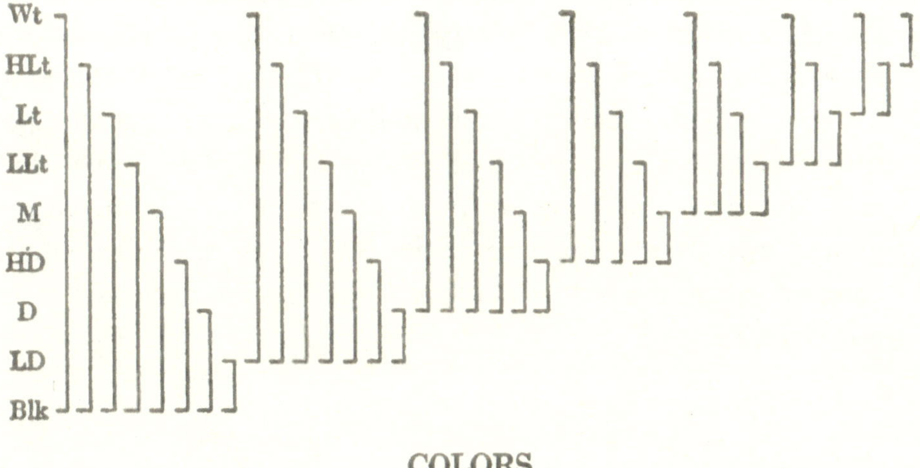


## COLORS

The quality of light in any tone, its color, may be Red (R), Yellow (Y) or Blue (B); or it may be Orange (O), Green (G) or Violet (V); or it may be one of the intermediates, Red-Orange (RO), Orange-Yellow (OY), Yellow-Green (YG), Green-Blue (GB), Blue-Violet (BV), Violet-Red (VR).

## THE SCALE OF COLORS

R RO O OY Y YG G GB B BV V. VR

The Scale of Colors appears in the Spectrum Band. The colors appear also in pigments of high intensity: in English Vermilion, Orange Vermilion, Cadmium Orange, Cadmium Yellow, Lemon Yellow, Vert Emeraude, Cerulean Blue, Cobalt Blue, and French Ultramarine, in Madder and Alizarin reds. They appear, also, in natural objects of high color intensity, in minerals and in precious stones, in leaves and flowers, in the wings of butterflies, and in the feathers of birds.

Any one of the twelve colors may be more or less neutralized or even completely neutralized by its complementary. The complete neutralization of any color gives us what we call Neutrality (N). The complementaries which neutralize and consequently balance one another are, approximately, Red and Green, Red-Orange and Green-Blue, Orange and Blue, Orange-Yellow and Blue-Violet, Yellow and Violet, Yellow-Green and Violet-Red, Green and Red, Green-Blue and Red-Orange, Blue and Orange, Blue-Violet and Orange-Yellow, Violet and Yellow, Violet-Red and Yellow-Green. These are the colors which, being mixed together, give us neutralizations and Neutrality. Complementary colors occur in the Scale of Colors at the interval of the seventh, approximately.

## COMPLEMENTARY COLORS

### AT THE INTERVAL OF THE SEVENTH

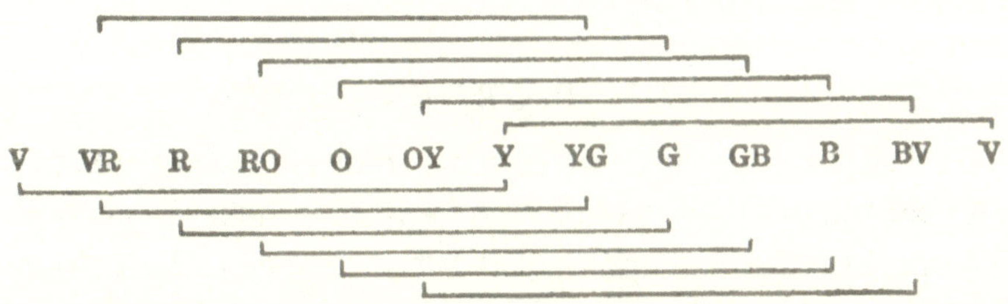


In saying that certain colors, Red and Green for example, are complementaries we must recognize the fact that these colors are variable under their terms or names. In pigments and on palettes, for instance, the complementaries are a particular red and a particular green which, being mixed, produce a colorless gray. As the Red varies towards Orange, the Green, to be a true complementary, must vary towards Blue. As the Green varies towards Yellow the Red must vary towards Violet. It follows that, having established on the palette a certain Red, the complementary will be the particular Green which will neutralize it; or having established on the palette a certain Green, its complementary will be the particular Red which will neutralize it; whatever that Red may be. The same is true of all complementaries: they must neutralize and so balance one another perfectly. We must be careful not to be influenced by the words or names we use and the effects which may be associated in our minds with those words or names. If, thinking of Red as the effect of English Vermilion and of Green as the color of green grass, we produce tones to express these ideas, we shall produce a Red and a Green which will not be complementaries. The mixture of a hot Red and a relatively warm Grass-Green will give us, not a colorless gray, but a red-brown or a green-brown, as we use more of the Red or more of the Green in the mixture. No colors are complementary on the Palette which do not, when mixed together, produce a perfect neutral. When the complementaries are unequally intense, one being stronger in color than the other, the neutral will be obtained by an unequal mixture: more Red than Green, for example, or more Green than Red.

### HOT AND COLD COLORS

Considering the different colors produced by pigments and pigment-mixtures we feel that some of them are relatively Hot (H) and others relatively Cold (C); that between the extremes there are colors which are half-hot and half-cold (HJC). The hottest of all colors is a Red-Orange, when it is pure and intense; as intense as possible. The coldest of all colors is a Green-Blue when it is as pure and as intense as possible. Violet is relatively half -hot and half -cold; Yellow is relatively half-hot and half-cold. Violet-Red and Orange-Yellow are, accordingly, in the first degree of heat; Yellow-Green and Blue-Violet in the first degree of cold. Red and Orange are in the second degree of heat; Green and Blue in the second degree of cold. Red-Orange, the hottest of all the colors, is in the third degree of heat and Green-Blue, the coldest of all colors, is in the third degree of cold. Neutrality, the result of mixing complementaries, is half-hot and half-cold, like Violet and like Yellow.

## The Colors in Pigments and Pigment-Mixtures

### RED (R)

Red, the color which we see in rubies, appears in a variety of pigments and pigment-mixtures. The mixture of Indian Red, Chinese or English Vermilion, and a little Madder or Alizarin Crimson gives a fine quality of Red. A touch of French Ultramarine is sometimes desirable. Red occurs in the value High Dark, approximately, and is neutralized by a Green of the same value, a Green which may be produced by a mixture of Vert Emeraude (Green Oxide of Chromium, Transparent) with a very small amount of Zinc White. Green is the clear cool color we see in the emerald; a color which cannot be produced by the mixture of Vert Emeraude with Lemon Yellow, the mixture commonly used. A mixture of Cerulean Blue with Lemon Yellow comes very near to it.

## RED-ORANGE (RO)

Red-Orange is seen in a high degree of intensity in the mixture of English or Chinese Vermilion and Orange Vermilion. Red-Orange occurs in the value Middle, approximately, and is neutralized by a Green-Blue of the same value, which may be produced by the pigment Cerulean Blue. A good imitation of Cerulean Blue may be produced by a mixture of French Ultramarine, Vert Emeraude and Zinc White.

## ORANGE (O)

The color Orange is well represented by a mixture of Orange Vermilion and Lemon Yellow. The color Orange occurs in the value Low Light, approximately, and is neutralized by Blue; a blue which may be produced by French Ultramarine or Cobalt with enough Zinc White to bring it to the value Low Light. If Cadmium Orange is used to represent Orange the Blue which will neutralize it must be a Violet-Blue. I rarely use either Cadmium Orange or Cadmium Yellow. They are too strong, too intense, to be satisfactorily neutralized by any Blue or Blue-Violet we are able to produce in corresponding values.

## ORANGE-YELLOW (OY)

The color Orange-Yellow may be produced by a mixture of Orange Vermilion and Lemon Yellow. Orange-Yellow occurs in the value Light, approximately, and is neutralized by a Blue-Violet in the same value; a Blue-Violet which may be produced by a mixture of French Ultramarine, Mars Violet or Indian Red, and Zinc White.

## YELLOW (Y)

The color Yellow occurs just as we want it in the pigment called Lemon Yellow (Barium Chromate). It occurs, also, in Strontian Yellow and in Zinc Yellow with a touch of White. A little White may be used with any of these yellow pigments, to give them a better consistency and a more delicate quality. Yellow occurs in the value High Light, approximately, and is neutralized by a Violet in the same value; a Violet which may be produced by a mixture of French Ultramarine and Mars Violet or Indian Red and Zinc White.

Madder or Alizarin Crimson might be used with French Ultramarine and Zinc White in producing Violet in High Light, but the Madder and Alizarin Reds fade out when used with a considerable quantity of White. For this reason it will be better to use Mars Violet or Indian Red. The Madder and Alizarin Reds are particularly good for glazing. They are not so satisfactory when they are used in mixtures with more solid pigments.

Cobalt Violet (Violet de Cobalt) might be used with Zinc White to represent Violet in High Light but in that case it would be necessary to give up the use of a steel palette knife. Cobalt Violet is a very fine pigment but its quality is destroyed by contact with iron.

## YELLOW-GREEN (YG)

The color Yellow-Green is seen in a mixture of Vert Emeraude with Barium, Strontian or Zinc Yellow, and a little Zinc White. Yellow-Green occurs in the value Light, approximately, and may be neutralized, in the same value, by a mixture of Mars Violet or Indian Red with French Ultramarine and Zinc White, enough White to pull the mixture up to the value Light. The neutralization of "Yellow-Green" by "Violet-Red" is often unsatisfactory; the result of the mixture being a warm brown instead of a neutral gray. That is because both the Yellow-Green and the Violet-Red are made too hot, too much yellow or too strong or too warm a yellow (OY) being used in the Yellow-Green, and too much red or too intense and hot a red in the Violet-Red. It must be remembered that Yellow-Green is a cool, not a hot color. It is one degree cold and Violet-Red is one degree hot, according to the theory on which we are proceeding.

## GREEN (G)

The color Green, which we should recognize in green jade and in emeralds, is seen in a mixture of Vert Emeraude with a little Zinc White. Vert Emeraude, which is the best pigment we have for greens, is so low in value that we are obliged to use White to pull it up to Low Light in which value its color is fully revealed. It must be remembered that Green is a cold color. It comes, in the scale, next to Green-Blue, the coldest of all colors. Green in its full intensity is seen in the value Low Light, approximately, and it is neutralized by a Red of the same value; a Red which may be produced by a mixture of English or Chinese Vermilion with a little Zinc White; enough White to raise the tone to Low Light.

## GREEN-BLUE (GB)

The color Green-Blue is seen, very nearly right, in the pigment called Cerulean Blue; or in a mixture of French Ultramarine and Vert Emeraude with Zinc White. Green-Blue occurs in the value Middle, approximately, and is neutralized by a Red-Orange in the same value; a Red-Orange which may be produced by a mixture of English or Chinese Vermilion with Orange Vermilion.

## BLUE (B)

The color Blue occurs, approximately, in the pigment called Cobalt. A light Cobalt is to be preferred to a dark Cobalt. The mixture of French Ultramarine with a little Vert Emeraude and White produces a good Blue but I prefer, when it is possible, to get the tones which represent the Spectrum Band, without using White. It is, of course, necessary to use White in in producing light tones of dark pigments. White, for example, has to be used with Vert Emeraude to produce Green in the value Low Light. Blue belongs in High Dark, approximately, and is neutralized and balanced by an Orange in the same value, an Orange which may be produced by a mixture of Burnt Sienna with a little Orange Vermilion.

### BLUE-VIOLET (BV)

The color Blue-Violet is seen in a mixture of French Ultramarine, Mars Violet or Indian Red, and Madder or Alizarin Crimson with a very little Zinc White to raise the value of the mixture and to reveal its color-quality. Blue-Violet occurs in the value Dark, approximately, and is neutralized by a tone of Orange-Yellow, in the same value, which may be produced by a mixture of Light or Venetian Red with Vert Emeraude.

### VIOLET (V)

The color Violet is seen in the mixture of French Ultramarine Blue, Mars Violet or Indian Red and a little Madder or Alizarin Crimson. Violet occurs in the value Low Dark, approximately, just above Black, and is neutralized by an extremely reduced Yellow, which may be produced by a mixture of Van Dyck Brown and Vert Emeraude. A little White may be used with these mixtures to raise the value and to reveal the color; I prefer, however, not to use White in producing dark tones. It seems to spoil their quality.

### VIOLET-RED (VR)

The color Violet-Red is seen in a mixture of Indian Red or Mars Violet with Madder or Alizarin Crimson and a little French Ultramarine. Violet-Red is not a hot color. It is as much cooler than Red as Red is cooler than Red-Orange. Violet-Red occurs in the value Dark, approximately. It is neutralized by a Yellow-Green which may be produced by a mixture of French Ultramarine and Vert Emeraude with just enough Yellow Ochre to turn it from Green-Blue to Yellow-Green.

The order in which the colors have been defined and described is the order in which they occur in the Spectrum Band; where the first color is clearly Red and the last color a suggestion of Violet-Red. The mixtures which I have proposed for the different colors and for their complementaries may be improved, no doubt; and other mixtures may be preferred, in view of particular effects to be produced. It is important, however, to have in mind a certain mixture and standard for each color and a mixture and standard for its complementary in the same value. After that, if there is any departure from the standard there must be a good reason for it.

The twelve colors, as we see and recognize them in Nature, vary indefinitely. They are sometimes bright and intense, sometimes dull and neutral. So on the palette. It is not always necessary or desirable to produce the colors of the palette in a high degree of intensity. It will often happen that a dull and neutral palette will serve the painter quite as well, if not better, than one in which the colors, being unnecessarily intense, must be constantly neutralized. Having no occasion to use intense colors, it is a good thing to neutralize all the colors of the palette before beginning to use them, or to use dull and neutral pigments in producing them. It will be desirable in many cases, to produce the Reds and Oranges of the palette without using Madders or Vermilion; using simply Mars Violet, Indian Red, Venetian Red or Light Red, and Ochre Yellows. The Madders and Vermilions should be used only when required. For Blues and Greens, it is often possible to use Blue Black and White, with a very small quantity of French Ultramarine or Vert Emeraude. The Old Masters used the mixture of Black and White for Blue constantly. The tones of the palette should in all cases be appropriate to the end in view. A neutral palette should be prepared for neutral effects.

## LIST OF PIGMENTS

The pigments which may be used in producing the tones of the palette are put down in the following list:-

Zinc White.

Rose Madder, or Alizarin Crimson; particularly good for glazing. Mars Violet. Indian Red.

English or Chinese Vermilion. Venetian or Light Red. Orange Vermilion. Burnt Sienna; very good for glazing. Cadmium Orange. Van Dyck Brown.

Raw Sienna or Mars Yellow; particularly good for glazing. Yellow Ochre. Cadmium Yellow. Aureolin; good for glazing.

Zinc Yellow.

Lemon Yellow (Barium Chromate), or Strontian Yellow.

Vert Emeraude (Green Oxide of Chromium, Transparent); good for glazing. Cerulean Blue. Cobalt Blue.

French Ultramarine Blue. Blue Black.

There are many good pigments which are not on this list. It contains only those which I am in the habit of using, which I know well, and am willing to recommend.

# The Palette, Brushes, Etc.

For a palette procure a good-sized sheet of transparent glass. Setting the glass upon a table with a piece of white cloth or white paper under it, you will have the best possible ground and surface for your tones; and it is a surface which you can keep clean, very easily. The advantage of having your tones on an even white surface will be at once appreciated. Accidental contrasts will be avoided and differences of value and of color in the tones more easily estimated.

In addition to the palette and the pigments a palette knife will be required and some brushes. The best palette knife will be a small one unless you are mixing up a great deal of paint. The brushes should not be too large or too small but suited to the purpose in every case; small brushes for small work, large brushes for large work. Some of them should be of sable, some of bristles; some flat and some round. The brushes should be washed out constantly, as fast as they get dirty, in a can of rectified turpentine kept alongside of the palette and conveniently within reach. The cans which have a grating upon which to rub the brushes when you are cleaning them are the best. The turpentine should be of superior quality as some of it will remain in the brushes and find its way into your painting. It is a good plan before putting the brushes away for the day to wash them with a good soap. This is easily done when the paint has already been washed off in the turpentine. Brushes kept clean and in good condition will last for a long time. As a rule, there is linseed oil enough in the pigments to make a good medium, with the small amount of turpentine which remains in the brushes when they are constantly washed out in it. If it is desirable to keep the tones on the palette from day to day, the palette, which is of glass, may be immersed in clear water. That means that a metal or china tray large enough to hold the glass palette will be required. When the palette is removed from the water it must be set up on end and the water drained off. When the palette is dry it may be necessary to moisten the tones with a little linseed oil and to soften them with a palette knife. Only the very best of linseed oil should be used.

## Theory of Tone-Relations

It appears from a study of the Spectrum Band and the Scale of Colors based upon it that there are two ways or modes of moving from Darkness into Light through the colors; one way is through Red up to Orange and from Orange up to Yellow; the other way is through Blue up to Green and from Green up to Yellow. The starting point of both of these color movements is in Violet; a Violet which is seen in certain pigments and in combinations of Red and Blue pigments. Violet may be regarded, therefore, as the least luminous of colors. Both movements terminate in Yellow which appears to be the most luminous of colors. Below Violet is colorless Darkness, represented in pigments by Black, and above Yellow is colorless Light, represented in pigments by White. The first of the two color movements, the Red-up-to-Orange movement, is relatively hot; the other, the Blue-up-to-Green movement, is relatively cold.

| THE HOT MOVEMENT UP TO LIGHT | | THE COLD MOVEMENT UP TO LIGHT |
|---|---|---|
| Wt | Wt | Wt |
| Y′ | HLt | Y |
| OY | Lt | YG |
| O | LLt | G |
| RO | M | GB |
| R | HD | B |
| VR | D | BV |
| V | LD | V |
| Blk | Blk | Blk |

The two movements up to Light, which I describe on facing page, one hot (H), the other cold (C), appear in the Spectrum Band in contrary motion. The hot movement meets the cold movement in the Yellow which is common to both; but the two movements, while they balance one another in this way, in contrary motion, are not in themselves balanced as they should be by the complementaries which are required for neutralizations. Yellow is not neutralized by Yellow, Orange-Yellow is not neutralized by Yellow-Green nor Orange by Green; Violet is not neutralized by Violet, Violet-Red is not neutralized by Blue-Violet nor is Red neutralized by Blue. It is only when we reach Red-Orange in one movement and Green-Blue in the other that we get the possibility of a neutralization and a balance. In order to neutralize and to balance the movement up to Light, which is hot, we have to set against it an inversion of the movement, which is cold. In order to neutralize and balance the movement up to Light, which is cold, we have to set against it an inversion of the movement, which is hot.

In the Spectrum Band Red is the initial color. The other colors follow in the order as described. Beyond Violet there is a suggestion of Violet-Red and of a return into the Light; so, we guess, a recurrence of Red and a repetition of the Band. It is a question, therefore, whether we should not, in our diagram, begin with Red, and, reading to the right through Orange, Yellow, Green, Blue and Violet come to an end in a Violet-Red. The reason for not doing this is found in the fact that the Violet-Red which appears at the end of the Spectrum Band has more light in it and is higher in its value than the preceding Violet. That means not only a return towards Red but a return from Darkness to Light. The colors, including Violet-Red, being set in the values in which they appear in the Spectrum Band, we get a very perfect system of Value and Color Balances; appealing equally to reason and to our love of Order and Harmony.

## THE THEORY OF TONE RELATIONS

| DARK | | | | | | LIGHT | | | | | | DARK |
|---|---|---|---|---|---|---|---|---|---|---|---|---|
| H½C | | | Cold | | | C½H | | | Hot | | | H½C |
| | | | | | | THE COMPLEMENTARIES | | | | | | |
| Y | YG | G | GB | B | BV | V | VR | R | RO | O | OY | Y |
| N | . | . | . | . | . | N | . | . | . | . | . | N |
| V | VR | R | RO | O | OY | Y | YG | G | GB | B | BV | V |
| | | | | | | THE SPECTRUM BAND | | | | | | |
| C½H | | | Hot | | | H½C | | | Cold | | | C½H |
| DARK | | | | | | LIGHT | | | | | | DARK |

It is upon this theory that the painter should prepare his palette; which will take the form indicated in the following diagram:

### THE COMPLETE PALETTE

### I HC

Based on the Theory of Tone Relations as Indicated in the Preceding Diagram

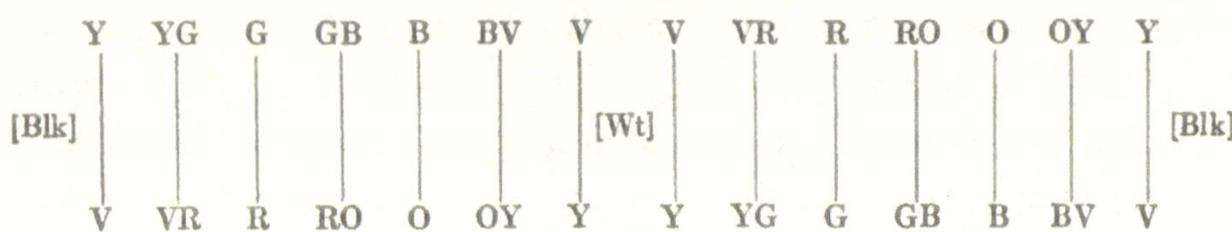


The tones of this palette may be used singly or in up-and-down mixtures of complementaries; in which mixtures, the colors, being complementary, may be more or less neutralized.

Theoretically, a half-and-half mixture of the complementaries should produce on the palette a line of perfect neutrals passing through the Scale of Values from Black, representing extreme darkness, up to White, representing extreme light, and from White down to Black. Theoretically, there is a central horizontal of perfect neutrality. As a matter of fact, however, the complementaries will rarely be so perfectly adjusted as to produce in a half-and-half mixture a perfect neutrality. Varying in intensity, one being more intense than the other, neutrality is likely to fall on one side or the other of the central horizontal. It will rarely happen that the line of neutrality will be exactly central or exactly horizontal. Using pigments, as they come, the painter will find it extremely difficult, if not impossible, to produce a perfectly adjusted and perfectly harmonized instrument; an instrument of precision; but he must try constantly to produce it, and he may possibly succeed. In the effort he may become himself an instrument of precision. Then, and not before, will he be able to produce a palette which will be one. Nothing expresses the master more clearly and unmistakably than his palette and the tones he produces upon it. Somebody has said, "Show me his palette and I will tell you whether he is a good painter or not."

The neutralizations and neutrals produced by the mixture of complementaries are far more interesting than any neutrals which can be produced by mixing colors with Black and White; or Black and White, the one with the other. The vibration of the complementary particles gives to these neutralizations, and even to what appears to be a perfect neutrality, an unmistakable liveliness. The tones of colors when mixed with Black and White and the tones of Black and White when they are mixed together are relatively dull and lifeless. The more or less neutral tones produced by the mixture of complementaries suggest, unmistakably, the effects of light and color which we see in Nature; which is a good and sufficient reason for preferring them.

The intermediate neutrals produced by the cross-mixing of complementaries make, with Black and White, a scale of neutral values which will be a far more interesting scale to use than any that can be produced by mixtures of black and white pigments. With the scale of colors neutralized by complementaries *grisaille* effects may be produced in which the play of complementary particles will give life and variety to the tones of gray and will suggest the colors which might have appeared had they been permitted to do so. The drawing and composition being established in *grisaille*, the *grisaille* being allowed to become dry and solid, we have a very good basis upon which to proceed to the coloring. In all serious undertakings this is the proper procedure; to get the drawing and composition settled in terms of neutrality before going on to the coloring; which is sufficiently difficult, when we get to it, to require all our interest and attention.

It is a serious objection to the mixtures of black and white pigments that the tones produced are not neutral. The use of black charcoal on white paper produces perfectly neutral tones, half hot and half cold, but no neutral tones can be obtained by the mixture of black and white pigments. The tones so produced suggest Blue or Violet unmistakably, and they have been used to represent those colors. Leonardo da Vinci speaks of the azure tones of Black and White. Instead of using Black with White to produce a scale of neutral values it is better to use a mixture of Black with Burnt Sienna, Van Dyck Brown, or some other dark brown pigment, which mixture, when combined with White, will give really neutral tones; half hot and half cold. The true and righteous neutrals, however, are those which are produced by the mixture of complementary colors in the values in which these colors naturally and properly occur.

Referring to the diagram of the Complete Palette (I HC): in case a tone below the value of Low Dark is required Black may be used. In case a tone above the value of High Light is required White may be used; and there is no objection, in practice, to the mixture of Black with Low Dark Yellow and Violet or to the mixture of High Light Violet and Yellow with White. As a rule, however, the mixture of High Light Violet and High Light Yellow gives a neutral high enough in value to represent and suggest White. When that is the case there is no occasion for using White.

In describing any tone it is necessary, first, to name the value of it, then the color of it, and lastly the degree of intensity in which the color appears. For example High Dark Red, one-half (HD-R½) means Red in the value High Dark, half intensity. Low Light Blue, three-quarters (LLt-B¾) means Blue in the value Low Light, three-quarters intensity. In this way we get a terminology for the tones of the palette which will serve our purpose when we have occasion to speak of them or to write about them.

The tones of the Complete Palette (I HC) may be mixed, not only on vertical but also on horizontal and on diagonal lines; the principle and rule of mixing complementaries only, being maintained; but somewhat less strictly. The lines upon which mixtures may be made are shown in the following Diagram:

PALETTE
I HC

WITH AN INDICATION OF POSSIBLE MIXTURES

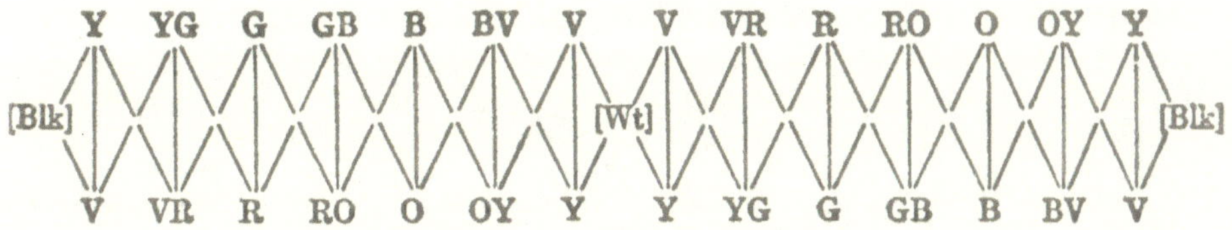


The tones connected by vertical lines are complementary. Mixtures between adjacents on horizontal lines of the same value range are complementary mixtures. Mixtures on the diagonals that cross one another are complementary mixtures. It is very important to keep the mixtures made on the different lines vertical, horizontal and diagonal distinct. There must be no confusion of these mixtures; because only those which are approximately complementary should be crossmixed or otherwise combined.

The tones of Palette I HC; and of the other palettes based upon it which will be described on the following pages, may be mixed, not only on vertical, horizontal and diagonal lines, but in triads; using, in each case, two tones which are complementary with a third tone which is adjacent to one of them. Red-Orange and Green-Blue, for example, may be mixed with either Red or Orange, or Green or Blue. The mixture of tones in triads will be described more particularly further on.

## Other and Simpler Palettes

Omitting some of the intermediates and their complementaries we get abbreviations of the Complete Palette which will, in most cases, serve the painter as well, if not better; the intermediates indicated by dots being obtainable, in somewhat diminished intensity, by mixtures on the line of dots.

In the following diagrams the reader will see what I mean when I speak of the abbreviations of the Complete Palette (I HC) produced by the omission of some of the complementary oppositions.

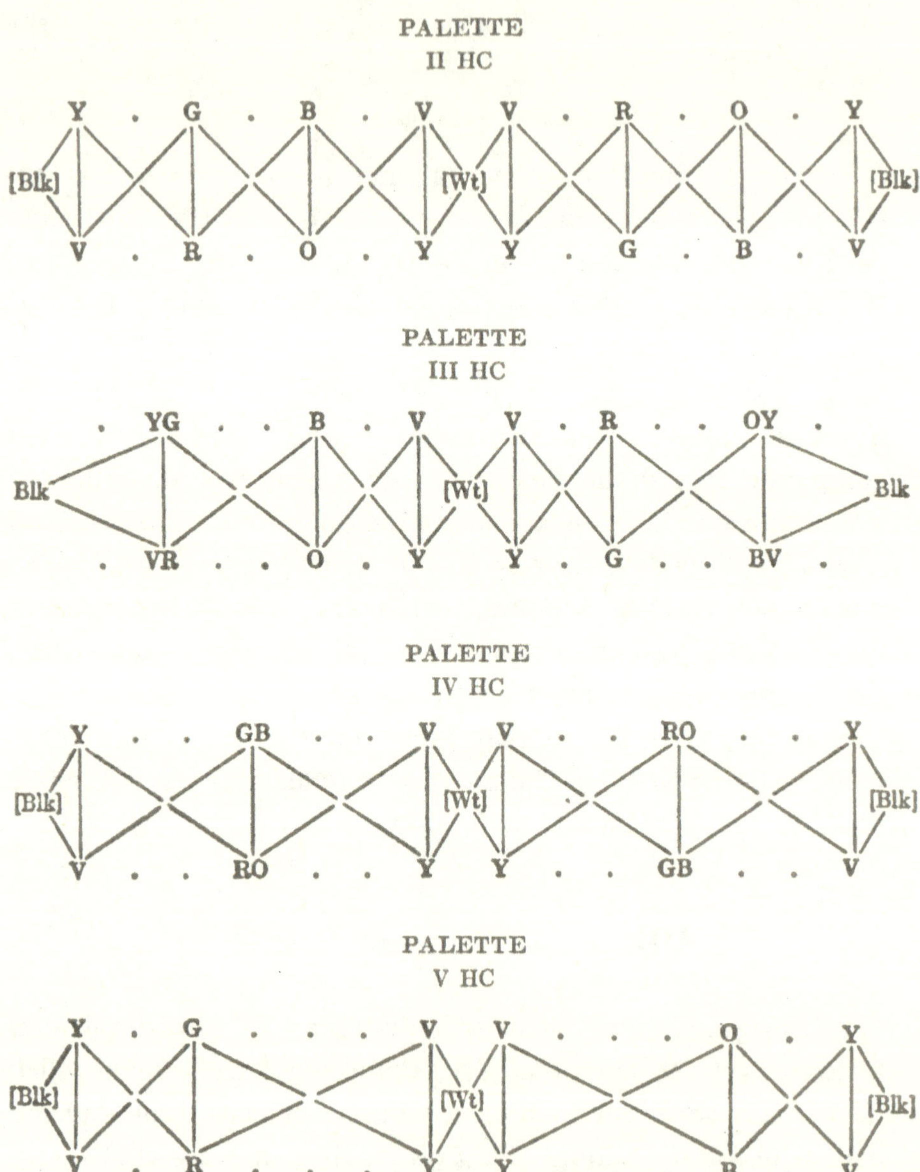

PALETTE
VI HC

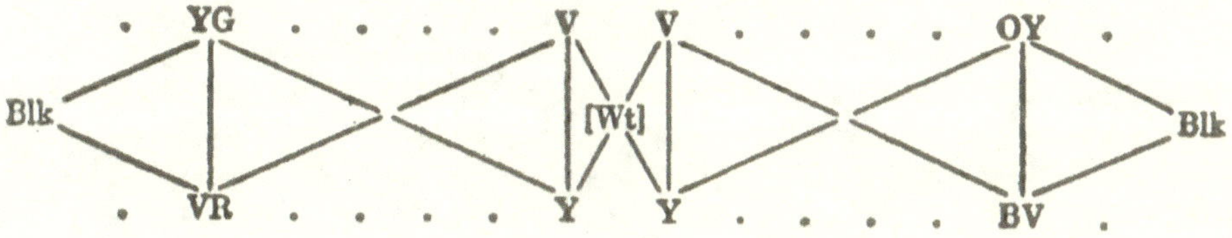

. YG . . . . . . V V . . . . . . OY .
Blk [Wt] Blk
. VR . . . . . . Y Y . . . . . . BV .

PALETTE
VII H½C

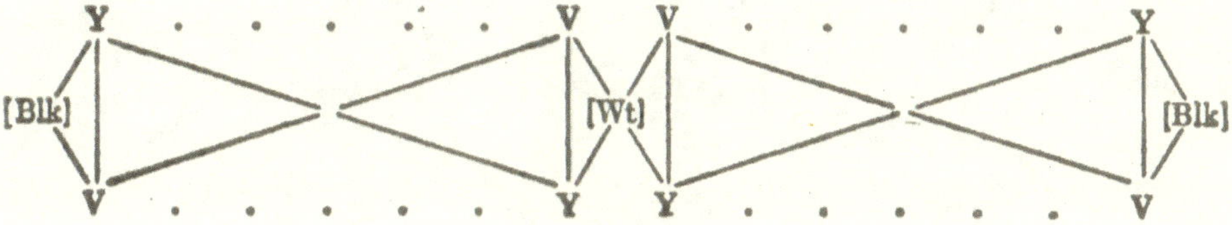

Y . . . . . . V V . . . . . . . Y
[Blk] [Wt] [Blk]
V . . . . . . Y Y . . . . . . . V

PALETTE
VIII HC

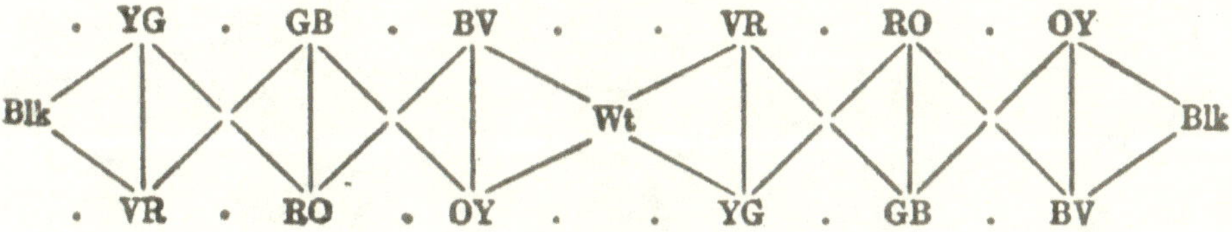

. YG . GB . BV . . VR . RO . OY
Blk Wt Blk
. VR . RO . OY . . YG . GB . BV

PALETTE
IX HC

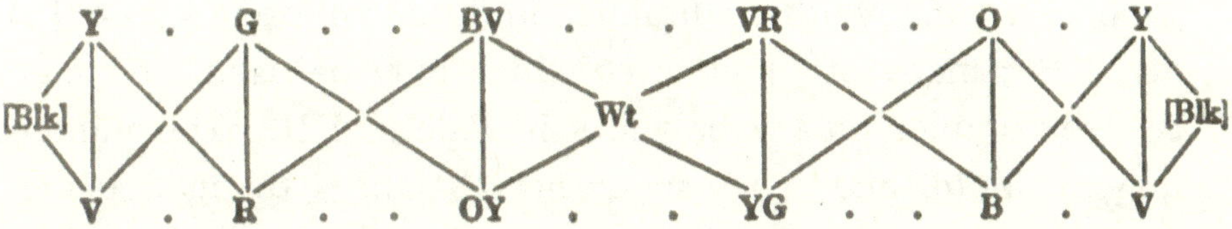

Y . G . . BV . . VR . . O . Y
[Blk] Wt [Blk]
V . R . . OY . . YG . . B . V

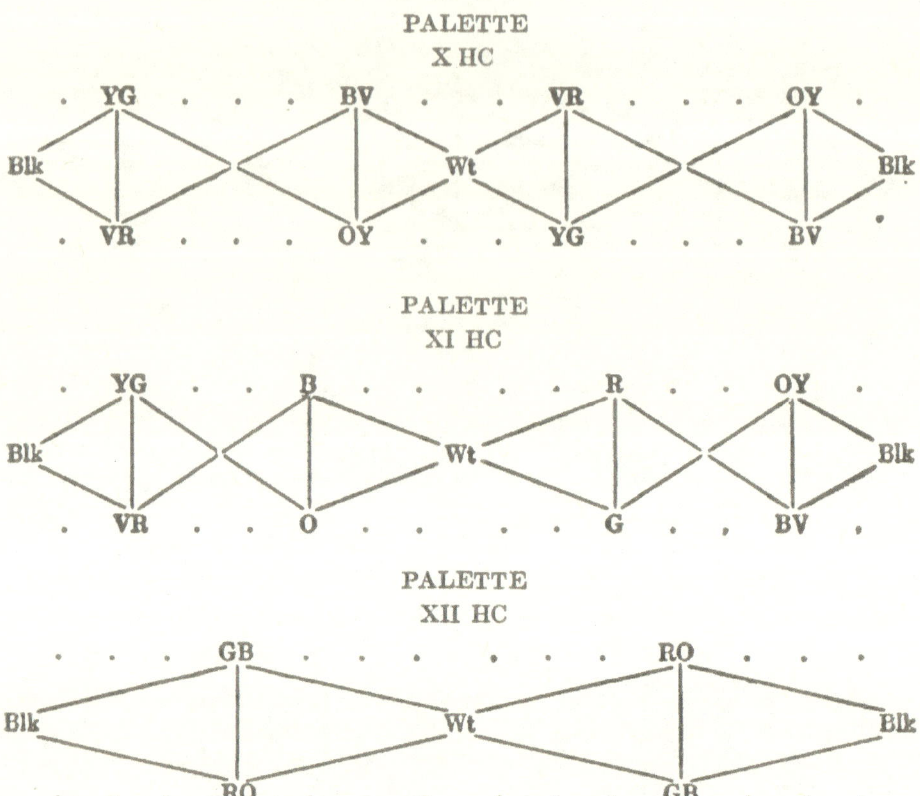


There are many other simplifications and abbreviations of Palette I HC; many other palettes which may be used. There are twelve complementary balances in Palette I HC. The number may be diminished to any six of these balances, to any five, four, three, two, or even one of them. A single balance of complementaries may be used as a palette with the pigments Black and White. With such a palette the expression of form is perfectly possible, and much more than that in some cases. Take, for example, Middle Red-Orange and Green-Blue with Black and White. With this palette (Palette XII HC) it is possible, not only to represent the forms of objects, to the last degree of specification, but to suggest the coloring; whether it is light or dark, hot or cold, intense or neutral; only local and particular colorings being missed. I am not recommending especially the palette of Middle

Red-Orange and Green-Blue with Black and White, but it is a palette with which consistent and interesting effects of light and color may be produced. I much prefer the effects produced by Palette IV HC in which Red-Orange and Green-Blue occur in Middle, and Yellow and Violet in Low Dark and in High Light.

It should be observed that as the diagonals are drawn at wider intervals of the Scale of Values they are drawn at shorter intervals of the Scale of Colors. On diagonals of the interval of the second in the Scale of Values we get a mixture of colors which in the Scale of Colors occur at intervals of the sixth. On diagonals of the interval of the third in the Scale of Values we get a mixture of colors which in the Scale of Colors occur at the intervals of the fifth. On the diagonals of the interval of the fourth in the Scale of Values we get a mixture of colors which in the Scale of Colors occur at the interval of the fourth. On the diagonals of the interval of the fifth in the Scale of Values we get a mixture of colors which in the Scale of Colors occur at the interval of the third. On the diagonals of the interval of the sixth in the Scale of Values we get a mixture of colors which in the Scale of Colors occur at the interval of the second. On the diagonals of the interval of the seventh we get the mixture of two tones of the same color. In view of these considerations it is evident that mixing the tones of the palette on the diagonals of different value intervals we get tones which would be impossible to get using only mixtures on the vertical and on the horizontal lines of the palette.

The pigments Black and White may be used in all of these palettes, either as Black and White or mixed; Black with the lowest tones of the palette, White with the highest tones. Where there is only one complementary balance on the palette, as in Palette XII HC, Black is used to pull the two tones, Red-Orange and Green-Blue for example, down into darkness, and White is used to pull them up to light. Instead of using black pigments which are relatively cold in tone, it is wise to neutralize the Black by mixing it with Van Dyck Brown or Burnt Sienna or some other dark brown pigment which will make the Black of the palette really neutral, half-hot and half-cold.

## Possible Combinations of Palettes

Any movement up to Light which is hot (H) may be combined in contrary motion, with any movement up to Light which is cold (C). The two movements of the palette are not necessarily symmetrical. For example; the H movement of Palette V may be set in contrary motion with the C movement of Palette III, with Low Dark Violet and Yellow substituted for Dark Blue-Violet and Orange-Yellow, as follows:

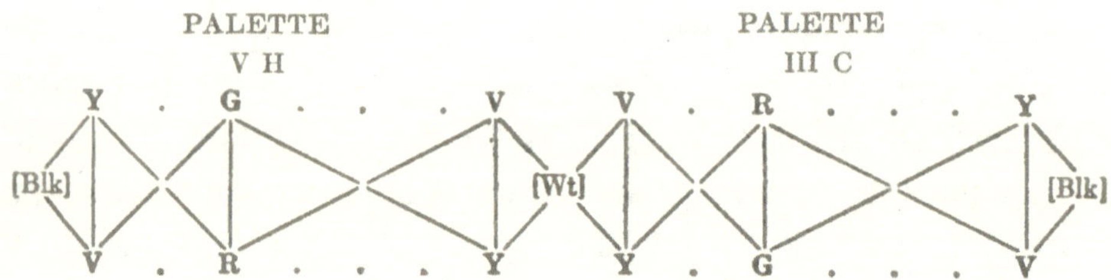


For another example; the hot movement of Palette IV HC may be set in contrary motion with the cold movement of Palette IX HC.

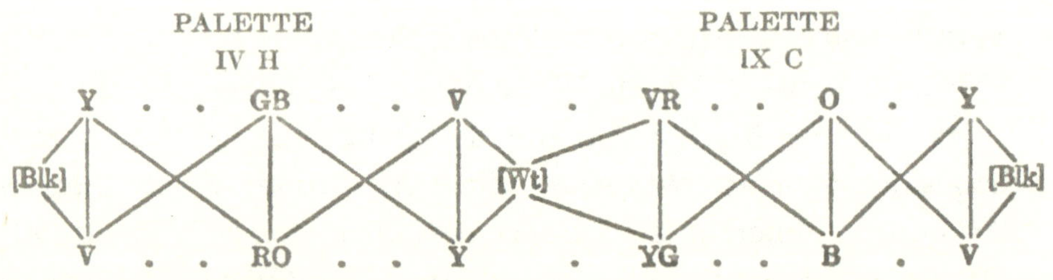


In this case the mixtures of White with High Light Violet and Light Violet-Red, and the mixtures of White with High Light Yellow and Light Yellow-Green are complementary, and may be cross-mixed or otherwise combined.

Any color movement up to Light which is hot (H) may be combined in contrary motion with any other movement up to Light which is hot (H). There is no objection to asymmetrical combinations of this description. For example:

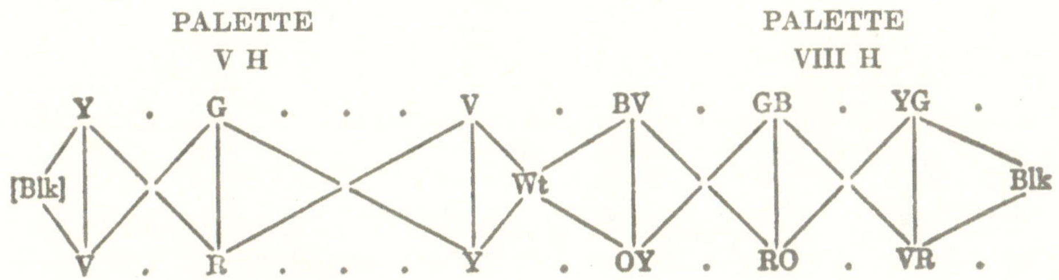


For another example:

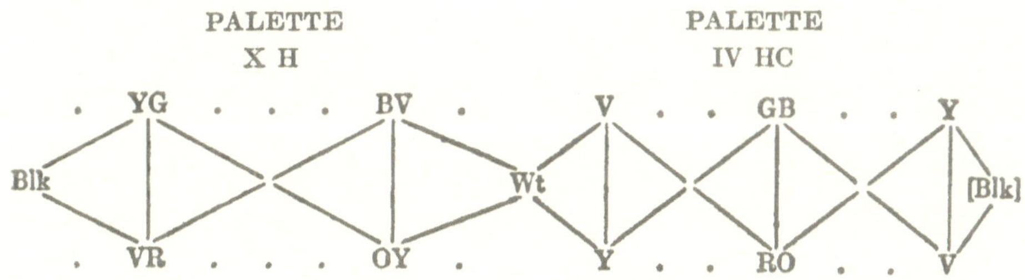


It is possible to connect the color movements of different palettes through Black, representing colorless darkness, as well as through White, representing colorless light; so that three or more palettes, - an indefinite number indeed, - may be strung together; some of them representing the movement up to Light which is hot, others the movement which is cold. Such a combination of color movements has a theoretic rather than a practical value. The painter is expected to use, in every case, the simplest palette which will serve his purpose. So doing, it will rarely happen that he will require on his palette more than two color movements; and it will more often happen that he will require only one.

## The Vertical Setting of Palettes

Instead of having two movements set in contrary motion, as in the palettes which I have been describing, the two movements may be separated, set vertically and used singly. We shall then have only one movement in each palette; either a movement up to Light which is hot or a movement up to Light which is cold.

| PALETTE I H | | | | PALETTE I C | | |
|---|---|---|---|---|---|---|
| | [Wt] | | Wt | | [Wt] | |
| V | . | Y | HLt | Y | . | V |
| BV | . | OY | Lt | YG | . | VR |
| B | . | O | LLt | G | . | R |
| GB | . | RO | M | GB | . | RO |
| G | . | R | HD | B | . | O |
| YG | . | VR | D | BV | . | OY |
| Y | . | V | LD | V | . | Y |
| | [Blk] | | Blk | | [Blk] | |

For the sake of convenience; to give space on the palette for each mixture so that there will be no confusion of mixtures; it will be well to set Palettes I H and I C, each in three columns; duplicating, in that case, the Yellow-up-to-Violet sequence of the complementaries. (See facing page.)

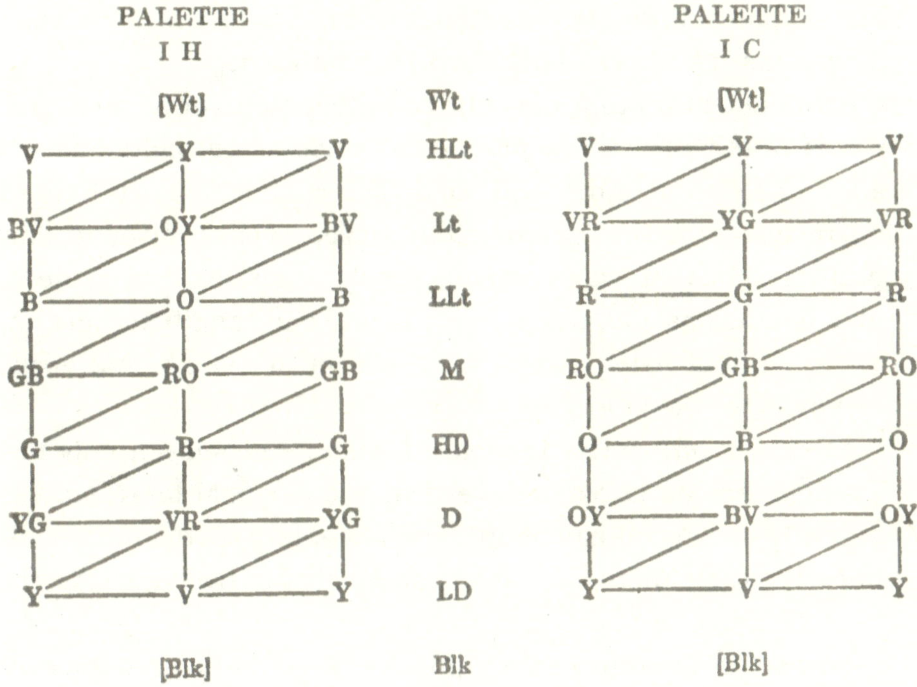


In Palette I H we have the movement up to Light which is hot (H) in the center, and an inversion of the movement up to Light which is cold (C) to the right and the left of it. In Palette I C we have the movement up to Light which is cold (C) in the center, and an inversion of the movement up to Light which is hot (H) to the right and the left of it. Lines have been drawn; horizontal, vertical, and diagonal to indicate possible mixtures. The tones on horizontal lines are complementaries. Mixtures on horizontal lines produce neutrals. Mixtures on opposite verticals, having the same range of values, are complementary mixtures. Mixtures on diagonals having the same range of values are also complementary mixtures. These diagonals cross one another in opposite directions when the palette is set in two columns. Being set in three columns the opposite and crossing diagonals become parallel. With the arrangement of these palettes in three columns it is possible to keep the mixtures perfectly distinct and to recognize the mixtures and tones which are complementary. This is very important as only those tones and mixtures which are complementary neutralize and balance one another.

Complementary tones may be combined without being mixed. A tone or mixture may be put on the canvas as a ground tone and allowed to become dry and solid. The complementary tone or mixture may be then scumbled or glazed over the ground thus prepared. In that way complementary tones may be made to play one against the other, neutralizing and modifying one another, without being mixed at all; except in the eye and visual feeling of the beholder.

The tones of Palettes I H and I C may be mixed not only one with another on horizontal, vertical and diagonal lines but also in triads, each triad having its complementary on the other side. In each palette there will be two sequences of twelve triads, the sequence on one side of the palette being complementary to the sequence on the other side. Opposite triads in these sequences will be complementary. Twelve (twice six) of these triads will be relatively hot and twelve (twice six) will be relatively cold. The triads which are complementary, hot and cold, are given the same numbers, as is shown in the following diagram.

It is possible to use the sequence of hot triads (1H, 2H, 3H, etc.) exclusively, or the sequence of cold triads (1C, 2C, 3C, etc.) exclusively, or the hot and cold triads in sequences of alternation (1C, 2H, 3C, etc. or 1H, 2C, 3H, etc.). It is possible, also, within the limitation of each triad, to move from hot tones up to cold tones, or from cold tones up to hot tones, or from lower neutrals up to higher colors, or from lower colors up to a higher neutrality. It is possible, also, within the limitations of each triad to establish not only sequences of gradation but also sequences of alternation between values, light and dark, and colors, intense and neutral; sequences in which the principle of Rhythm will be felt and observed. In other words the Palette becomes an instrument with which various forms of Pure Design, the repetitions, the sequences (progressive or rhythmical) and the balances (axial or radial, obvious or occult) may be simply and easily achieved. In this way the Palette serves the painter and designer both as a mode of thought and an instrument of expression.

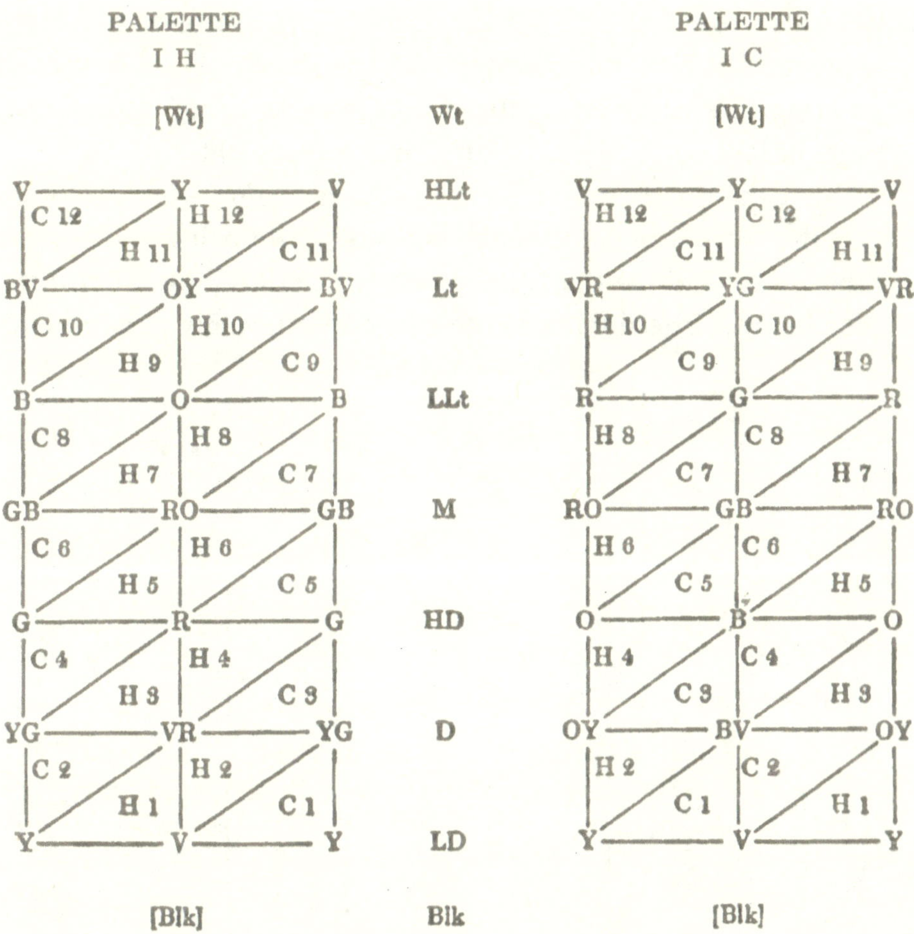


Palettes I H and I C may be used in still another way, the idea being to get sequences of triads which will be complementary one to the other; and triads in these sequences in each one of which there will, also, be a relation of complementaries. (See below.)

Each of the Palettes I H and I C is in this way divided into two palettes which may be used separately or both together; according to the end in view.

The composition being established in grisaille, using Black, the neutrals produced by a straight-across mixture of complementaries and White, the coloring may be developed in a second painting; the color elements of the grisaille being brought out more or less distinctly, in coloring more or less intense.

Using the three-column form in setting the palettes and omitting the two-letter intermediates, we get two palettes of twelve tones each: four of the tones in each palette being repetitions. (See below.) These two palettes may be especially recommended.

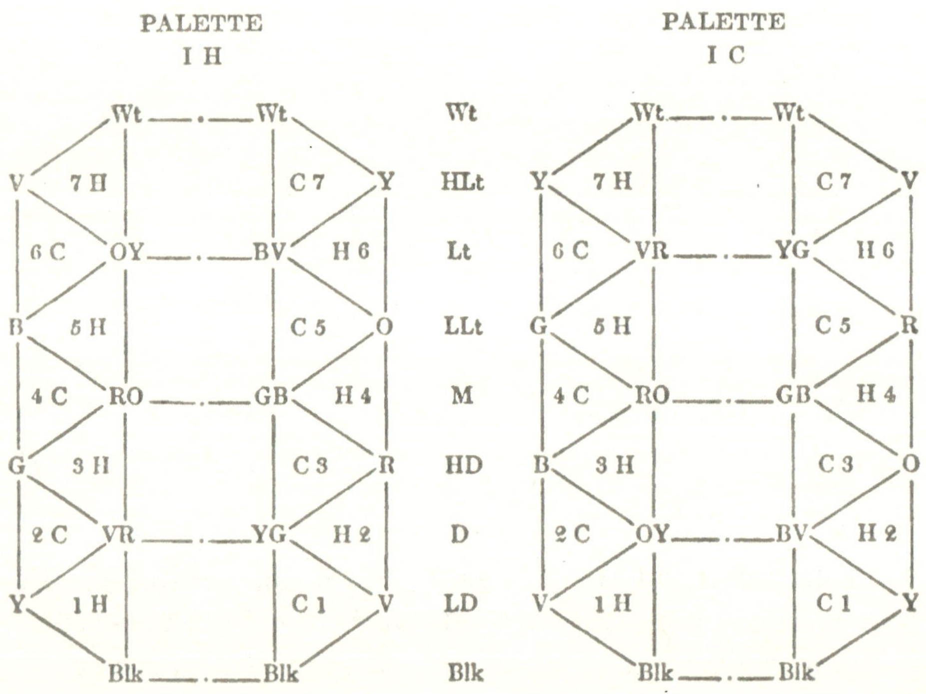

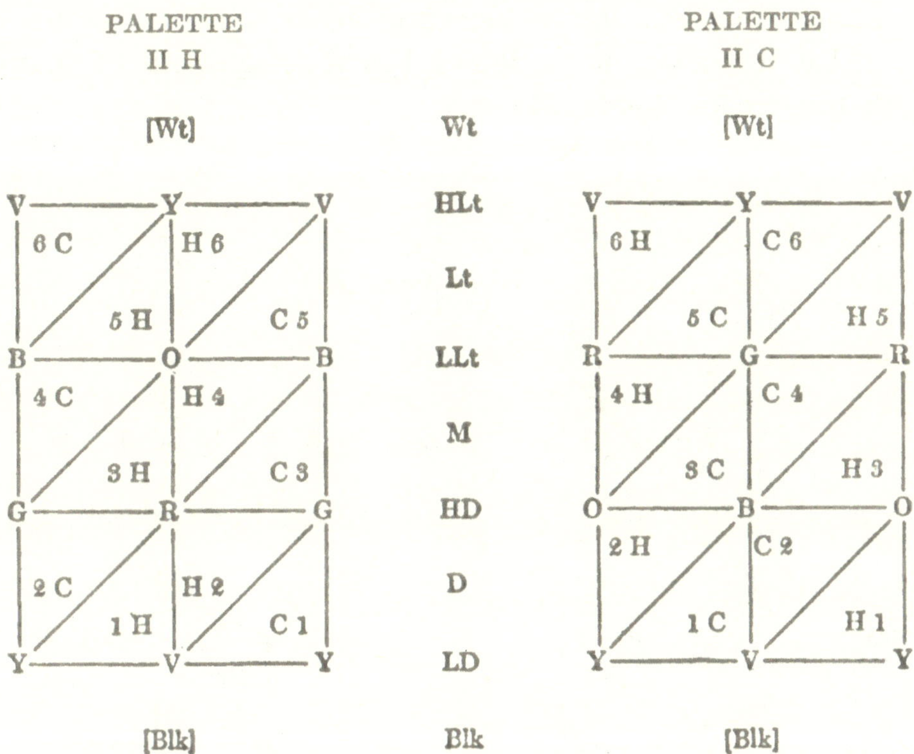


Mixing the tones on vertical and diagonal as well as on horizontal lines and in triads, these palettes of eight tones each, four of them being repeated, will serve the painter, for most purposes, quite as well as Palettes I H and I C, with fourteen tones each, seven of them being repeated.

So with the other palettes. Each one may be divided into two, set vertically and in three columns.

In using Palette III H (see next page), the painter will find light greens between Blue and Yellow. He can warm them by adding more or less of the Orange, or he can cool them by going up to the Violet. He will find dark greens between Yellow-Green, Orange and Blue. For flesh tones in light he can use the triad Orange, Yellow, Violet. For flesh tones in shadow he can use the triad Violet-Red, Orange, Blue. In Palette III C, howeter. he will get light greens which will be more luminous and flesh tones which will be more brilliant; having Red and Green in Low Light instead of Blue and Orange. He can, of course, use both palettes, if he has occasion to do so. He will then be using Palette III HC, set in three columns instead of two.

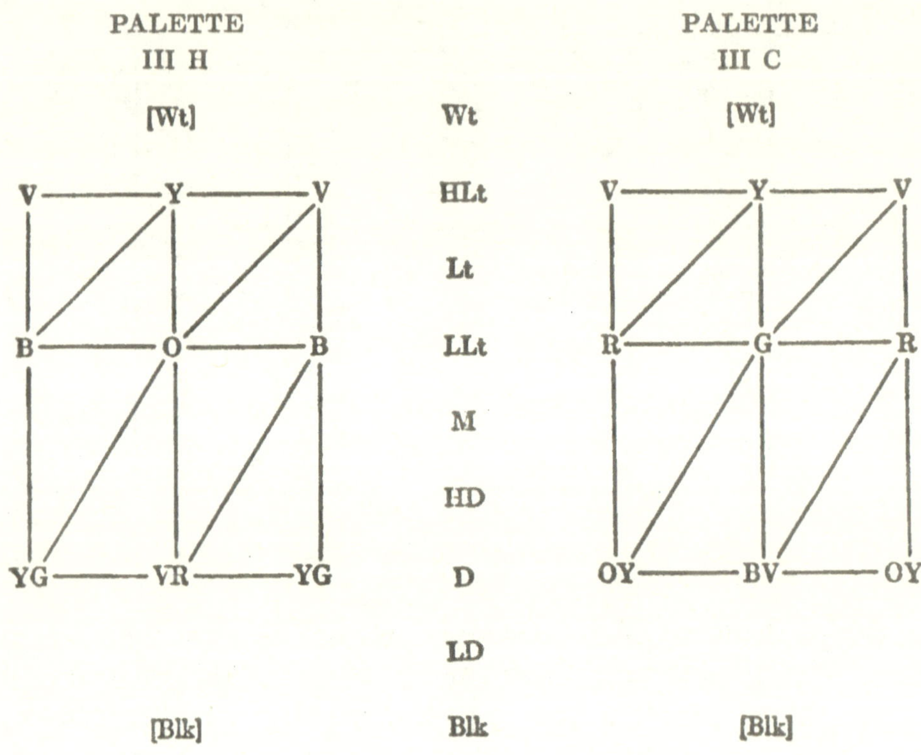

PALETTE
IV HC
Set vertically

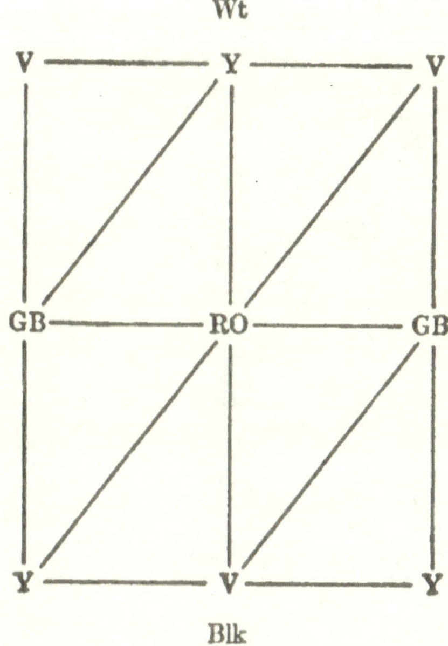


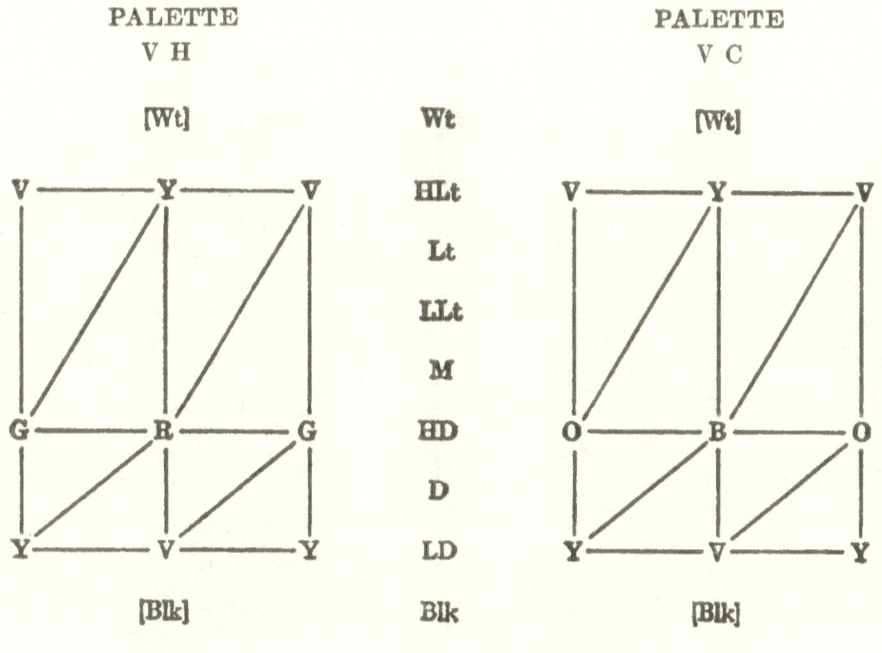

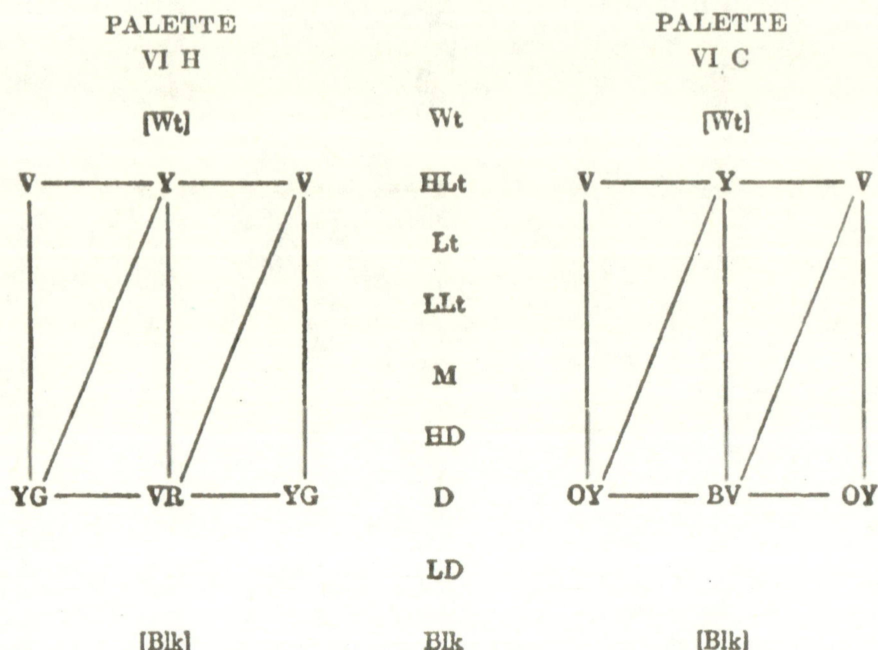


PALETTE
VII H½C
Set vertically

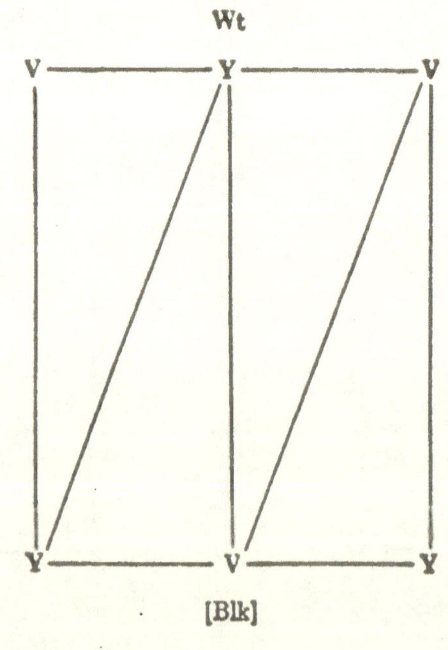

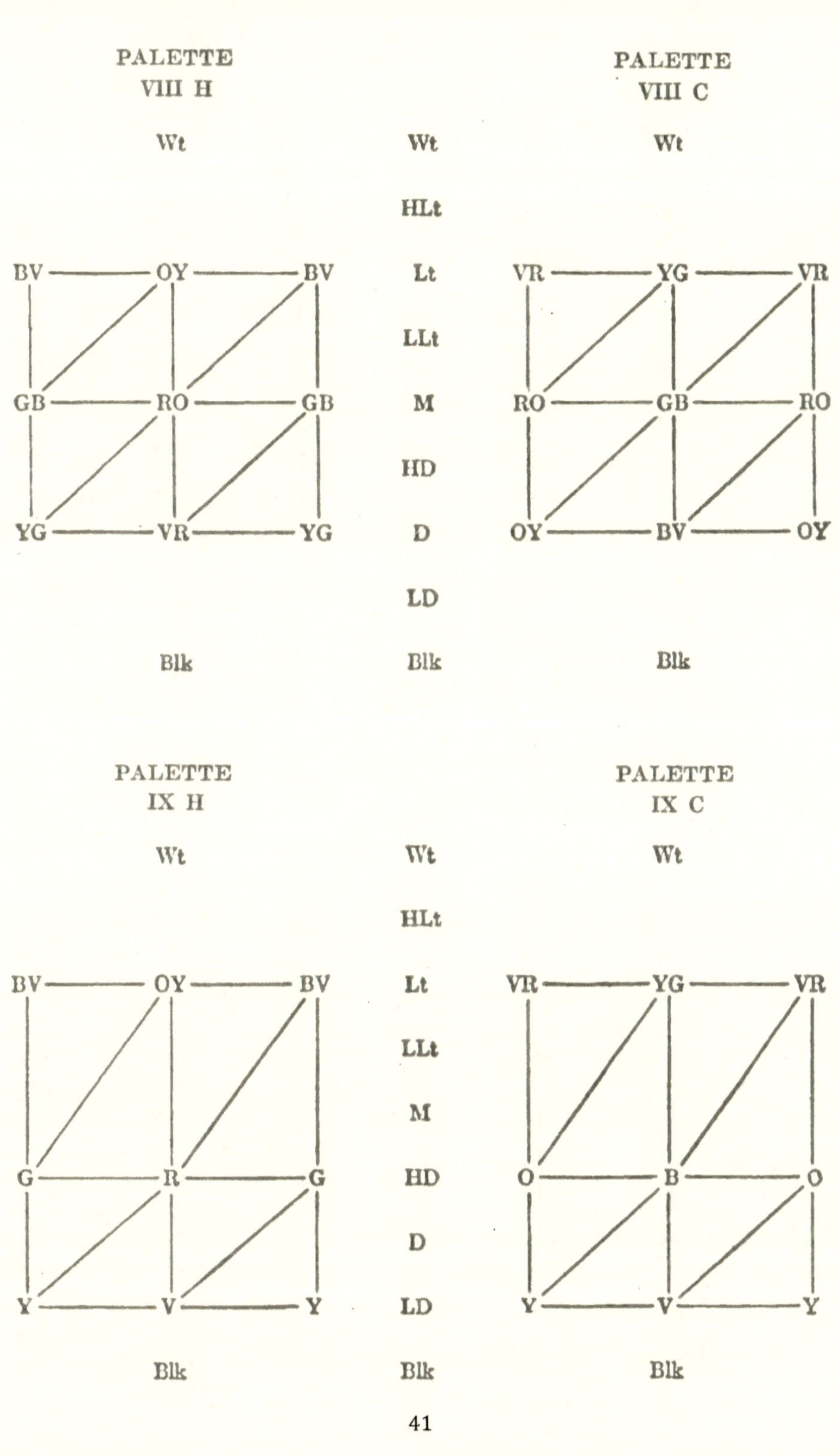
PALETTE
VIII H
PALETTE
VIII C
Wt
Wt
Wt
HLt
BV OY BV
Lt
VR YG VR
LLt
GB RO GB
M
RO GB RO
HD
YG VR YG
D
OY BV OY
LD
Blk
Blk
Blk
PALETTE
IX H
PALETTE
IX C
Wt
Wt
Wt
HLt
BV OY BV
Lt
VR YG VR
LLt
M
G R G
HD
O B O
D
Y V Y
LD
Y V Y
Blk
Blk
Blk

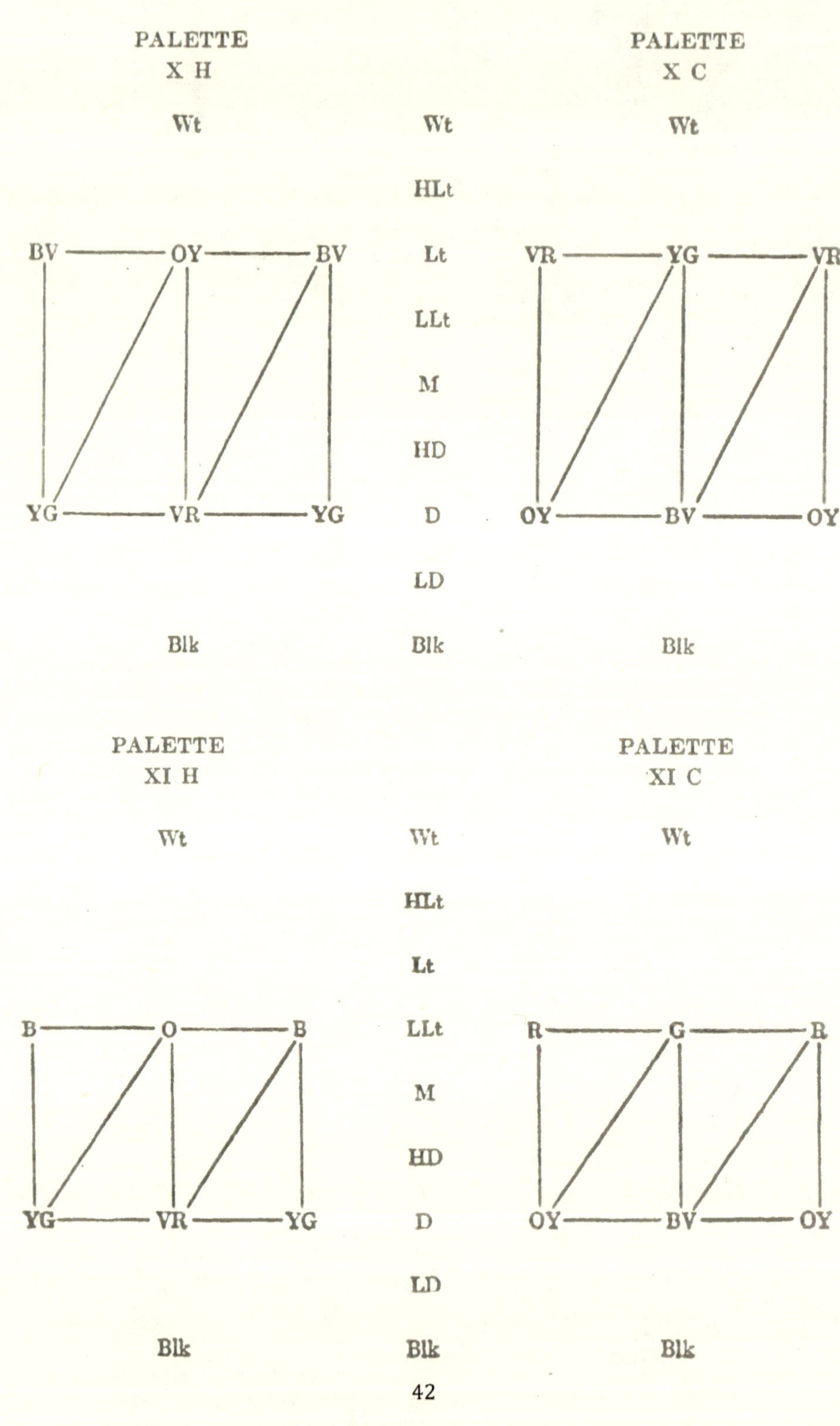

PALETTE
X H
PALETTE
X C
Wt
Wt
Wt
HLt
BV
OY
BV
Lt
VR
YG
VR
LLt
M
HD
YG
VR
YG
D
OY
BV
OY
LD
Blk
Blk
Blk
PALETTE
XI H
PALETTE
XI C
Wt
Wt
Wt
HLt
Lt
B
O
B
LLt
R
G
R
M
HD
YG
VR
YG
D
OY
BV
OY
LD
Blk
Blk
Blk

PALETTE

XII HC

Set vertically

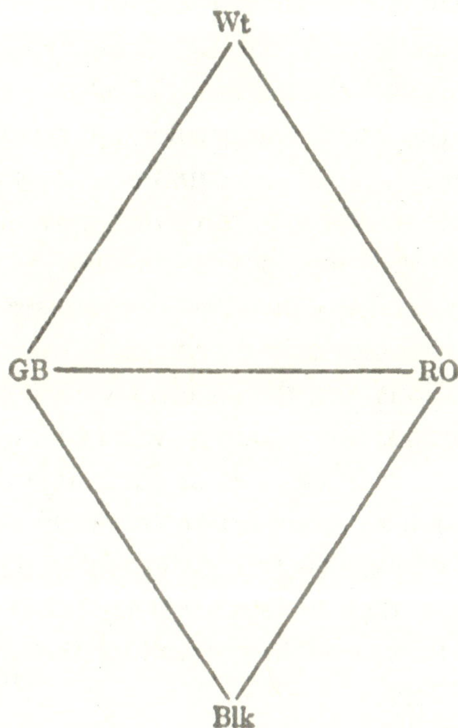


Using any one of the H series of palettes and finding it impossible to get certain tones that are required, the painter is at liberty to use in connection with it its complement in the C series: or using any palette in the C series and finding it impossible to get certain tones that are required it is possible to use, in connection with it, its complement in the H series. This means using an HC palette instead of an H or a C palette.

I prefer, as a rule, the vertical setting of these palettes. Instead of using two or more color movements set horizontally, end to end, I use them set vertically, side by side, as separate palettes. I then use one or the other, as occasion may require.

If pastels, crayons and colored pencils were produced representing the tones of any of these palettes, Palette II HC, for example, we should have a very simple means of studying the phenomena of light and color and a means of making, easily and quickly, records of visual experience and schemes of design. With such a mode of thought and expression the study of light and color might be introduced into schools and colleges where the use of palettes with the materials of oil-painting is difficult, if not impossible. The use of water colors, in accordance with the system of complementaries and complementary mixtures, is difficult. It is only the master who can paint in water colors without making a mess of it. Using pastels, crayons and colored pencils the student will see clearly the effect of complementary mixtures and will come to recognize and appreciate the value of the system of tone-relations and the palettes which I am describing. When he understands the system he will be able by practice and in due time to use it in water color painting, in which fine discriminations in light and color are so very difficult to achieve.

With pieces of white glass and of glass colored according to the tones of any one of these palettes, Palette II HC for example, we should have the best possible material for the production of designs in Stained Glass.

Having used all of these palettes, and others arranged on the same principles, and having compared the effects of light and color produced (for the most part in figure and portrait work), I find that very different effects and tonalities are produced by different palettes. This is due, first, to the tones of the palette, which may be more or less luminous or more or less intense. It is due, also, to the differences of value and color intervals. Using Palette I HC it is possible to achieve the effect of many colors in a high degree of intensity in a very bright light. This effect cannot be produced, in the same degree, with any other palette. The effects produced with Palette II HC will be less brilliant, though the effort be made to neutralize the tones of the palette as little as possible. With Palette III HC the tones between Low Light and High Light will be fairly intense and brilliant; but the tones in the lower values will be relatively neutral and dull. With Palette VIII HC, intense, brilliant tones can be produced only between the values Dark and Light. With Palette IV HC the most intense and brilliant coloring will be in the value Middle and just above and below it. The Yellow in High Light, though brilliant in itself, is relatively neutral, that is to say, half-hot and half-cold, and no very strong contrast of color can be produced between this Yellow and the Violet, its complementary, which is also half -hot and half -cold. The effects, therefore, which are produced with Palette IV HC will be decidedly neutral, and the effects produced with Palette VII HJC will be still more neutral. With a full and complete palette in the lights effects of high luminosity and brilliancy may be produced; with a full and complete palette in the darks the effects will be of low light. The Reds and Blues may be intense, but Red and Blue are colors of low degrees of luminosity. The painter must know what he wants and he must set his palette accordingly.

As in Music we pass from one key to another, modulating through notes and chords which are common to both keys; so in Painting, it is possible to modulate from one palette to another, from one tonality to another, making the transition through tones and mixtures which are common to both palettes. It must be remembered, however, that in Music the sounds of one key have passed out of hearing before those of another can be heard, whereas, in Painting the tonality of one palette and the tonality of another may be seen and felt simultaneously. In my own experience, it rarely happens that I feel the need of more than one palette. I generally use either one of the HC series, or one of the H series, or one of the C series. I rarely use any asymmetrical forms of the HC series. There is, really, no occasion to use more than one palette in any design or picture unless you are trying to imitate or copy what you see in all of its particulars. This may be the purpose of the painter but it is rarely the purpose of the artist; who rejoices in consistency and harmony and is less interested in what may be called the particulars or statistics of vision. This interest in the particulars of vision belongs rather to Science than to Art and is irrelevant to one very important branch of Science, the Science which means a knowledge and appreciation of Order and the Beautiful. The aim of the painter who is an artist is never imitation. His object is not to copy but to represent. The truth he achieves lies in a relativity of tones producing a true effect; an effect which can rarely, if ever, be achieved by the imitation and copying of particulars observed.

It is a good rule for the beginner and student to use one palette exclusively, say Palette II H or II C or the combination of these two in II HC; until he gets a clear knowledge of its limitations and its possibilities. In becoming master of one palette he is in the way of becoming a master of all of them. They rest, all of them, on the same principle and are used in the same ways.

No two palettes will give quite the same relativities and effects. To understand this, the student should analyze and represent a certain effect of light using a number and variety of palettes; observing how the truth of form may be expressed in different terms, in different effects, in different tonalities. Take any effect in Nature; the effect of a certain model; in a certain pose, in a certain place, in a certain light; with certain accessories, clothes, etc. Change the light from diffused daylight to concentrated daylight; from daylight to electric light; from cold electric light to hot electric light; use mirrors, one, two or more, to vary the reflections, and you will understand what it means to reveal the truth of form in different effects of light and color; in the different tonalities of different palettes. What Nature does is what the painter is supposed to do, following her example. Take certain pictures and designs, preferably the work of the great masters, and render them in the different tonalities of different palettes. It is a most interesting study and the best possible discipline and training. By this training the painter will come to understand, at' last, how his object is, not so much to copy and reproduce what he may happen to see at a certain moment and in a certain place; not so much that as to follow Nature in her principles and laws, and in so doing to produce results and effects which will be comparable with hers in their lawfulness and consistency. Whether they exactly correspond with hers or not becomes a secondary consideration.

Each palette has its limitations and its possibilities. What the limitations are is easy to see and understand after some experimental practice, but the possibilities are indefinite and infinite. The result and effect depend, not only upon the tones of the palette and upon the mixtures which may be made of those tones, but, still more, upon the positions, measures and shapes which are given to the tones in the design or picture which is being produced. Definite as they are in themselves, these palettes are infinitely suggestive and perfectly inconclusive. They are simply a metrical system in tone-relations.

The painter will, of course, choose his palette from the HC series or from the H series, or from the C series, or choose some combination of palettes with a view to the effect or effects of light and color he wishes to produce. As a rule he will choose one or another of the abbreviations of Palette I HC. The use of the complete palette or even the half of it, I H or I C means keeping a very large number of tones carefully prepared in their proper values, colors, and intensities, and in a condition ready for use; and this unnecessarily. Why use Palette I HC when you are able, using Palette IV HC, to produce all the tones you require? Why use Palette I H or I C when VI HC will give you all the tones you need. The painter should choose one palette rather than a combination and among the palettes the simplest one that will serve his purpose.

If the painter is interested only in form and not in color he should use Palette VII H½C and produce his composition and drawing in *grisaille.* Being interested in space-relations and the expression of form he should follow and express his interest in the simplest mode that is appropriate for the purpose. Coloring, if uninteresting, should be left out; it complicates the result and makes an irrelevant appeal. In Art we are supposed to express our interests, whatever they are; with the omission of all that is inconsistent with them or irrelevant to them. If the painter is interested in color relations mainly, the associated forms being relatively uninteresting, it is unnecessary for him to do more than reproduce the effect of light and color; a careful delineation of the forms which do not interest him being irrelevant to his purpose. It is a mistake to feel obliged, when you have an idea to express, to express other ideas which do not interest you; just for the sake of what is conventionally regarded as completeness. Of course, if your interest is completeness it must be achieved. The rule is to study what interests you and to express what interests you. In that way your art becomes the expression and measure of your mind and character. Your interests arid ideas being set forth and exposed, you are judged, praised or condemned accordingly.

Apart from interests and ideas, is the Art of Painting; with its materials, its modes, its methods, and its laws. The Art of Painting has been very slowly developed and established, by good precedents, and by the exercise of reason and common sense in connection with them. In ignoring or disregarding the Art so established; in using strange materials; in following unprecedented methods and modes; in disobeying the rules and laws of the Art; you are not expressing yourself in the Art but experimenting with it; perhaps with a view to changing it. That may or may not be worth while. In the meantime the Art of Painting is the same for all painters, for all artists. It is like the language which we all use. The question is: what have you or I to express by Painting; what are your interests; what are my interests; what ideas have you to express; what ideas have I to express? In expressing ourselves, our interests, our ideas, we use the same Art, you and I.

Painting means taking tones from the palette and giving them positions, measures, and shapes on some surface upon which we are producing a design or picture. Given certain tones; the positions, measures, and shapes are infinitely variable. Given certain positions, measures, and shapes; the tones are infinitely variable. It is a question of tones and it is a question of positions, measures, and shapes. The tones depend upon the palette and the method of using it. The positions, measures, and shapes which are given to the tones depend upon visual experience and the imagination. We may use the same palette and follow the same method. There is nothing personal or exclusive about the palette; but vision is personal, and the imagination is even more personal than vision. We are not judged by the palette which we use or by the method we follow in using it, but rather by what we see, imagine, and express. The piano is an admirable instrument. There are many players, but very few artists. We may all of us use the same palette in the same way, and only one of us produce anything that is interesting or significant. We may all be painters and there may be only one artist among us. We call him the artist not because he uses a palette but because he has a good reason and occasion for using one.

It must be understood that these set-palettes are not proposed as recipes for Truth or the Beautiful. They are simply modes of thought, in which we are able to think in definite terms to definite ends. They are, also, instruments which we may use in expressing ourselves.

Whether the painter is a person of knowledge and understanding, of intelligence and ability or not, whether he has good judgment and good taste or not, makes all the difference in the world. Unintelligent and stupid people will produce unintelligent and stupid paintings. There is no help for stupidity. It is bound to express itself in spite of the palette. The greatest fool I ever met was a master of four languages in each of which he expressed himself, unmistakably. The palette is merely a mode and means of expression; with its laws and its rules which are its grammar. It is the same in painting as it is in speech and in writing. The speaker or writer must have more to offer to us than a knowledge of the language. What is it that he has to say? What is it that he will put in writing? Back of the language is the man who uses it. Behind the palette is the painter; and the question is, always: has he anything to express by painting? If he has something to express; something interesting and significant; a properly set palette will be of the greatest value to him. When he has mastered it and knows how to use it, he will say of it that he cannot express himself without it; which is perfectly true.

The painters of the Middle Ages and of the Renaissance used carefully set palettes and definite tone-relations. This is proved by the recurrence, again and again, of exactly the same tonalities and effects. Modern painters have, as a rule, avoided the use of set-palettes and tone-systems; preferring to depend on visual feeling or native genius. In so doing they have made a very great mistake, and some of them are now fully aware of this. There is no getting on properly and successfully in any art without metrical systems or modes, in which it is possible to think definitely and to express oneself in what will be recognized as good form. There is, indeed, no art which can be satisfactorily and successfully practiced without constant reference and obedience to mathematical principles, systems, and laws. Through order we come to the beautiful. This idea, which I have had in mind as the basis of my teaching, is particularly well expressed by Poincare in the introduction to his book on the Value of Science; where he says: "What we call objective reality is, in the last analysis, what is common to many thinking beings, and could be common to all; this common part, we shall see, can only be the harmony expressed by mathematical laws. It is this harmony then which is the sole objective reality, the only truth we can attain; and when I add that the universal harmony of the world is the source of all beauty, it will be understood what price we should attach to the slow and difficult progress which little by little enables us to know it better."

www.ingramcontent.com/pod-product-compliance
Lightning Source LLC
LaVergne TN
LVHW050946080826
845145LV00004B/1435

*9781789875157*